CASTOR OIL!

Its Healing Properties

A Health Learning Handbook

Beth M. Ley, Ph.D.

——————— **BL Publications** ———————
Hanover, MN

Copyright © 1989, 1996, 2003. All rights reserved. No part of this book may be reproduced or transmitted by any means, electronic or mechanical, without written permission from the publisher:

BL Publications, Hanover, MN
1-877-BOOKS11
www.blpublications.com
email: bley@blpublications.com

Library of Congress Cataloging-in-Publication Data
Ley, Beth M., 1964-
Castor Oil! Its Healing Properties /Beth M. Ley.
    p. cm. -- (A health learning handbook)
ISBN: 1-8990766-27-5

Printed in the United States of America

This book is not intended as medical advice. Its purpose is solely educational. Please consult your healthcare professional for all health problems.

Cover photo by permission: larvalbug.com

YOU NEED TO KNOW...
THE HEALTH MESSAGE

Do you not know that you are God's temple and that God's Spirit dwells in you? If anyone destroys God's temple, God will destroy him, For God's temple is holy and that temple you are. *1 Cor. 3:16-17*

So, whether you eat or drink, or whatever you do, do all to the glory of God. *1 Cor. 10:31*

Table of Contents

Introduction 4
Dr. William McGarey Interview: 6
Castor Oil: What Is It? 8
Castor Oil: Its Toxicity .10
Castor Oil: What It Can Do 11
Castor Oil and Olive Oil 22
Castor Oil: How To Apply 23
Castor Oil: How It Works 25
Bibliography 26

Introduction

Health Learning Handbooks are designed to provide useful and interesting information about ways to improve health and well-being. Education about what good health is and what the body needs to obtain is crucial to obtain and maintain good health.

Good health should not be thought of as absence of disease. We should avoid negative disease-orientated thinking and try to concentrate on what we have to do to remain healthy. Health is maintaining on a daily basis what is essential to the body while disease is the result of attempting to live without what the body needs. We are responsible for our own health, and should get in control of it. If we are in control of our health, disease will not take control.

Our health depends on education. *My people perish for lack of knowledge.* (Hosea 4:6)

This book discusses castor oil and its healing properties. The information in this book is largely taken from Dr. William McGarey's book, *Edgar Cayce and the Palma Christi.* Dr. McGarey, Chairman of the Board of the A.R.E. Clinic in Phoenix, AZ, has a vast amount of clinical experience using castor oil and with tremendous success. In the course of his medical career spanning over several decades, Dr. McGarey has published numerous articles and books covering treatments with various Cayce remedies. He also has a revised and updated book about castor oil, *The Oil That Heals.*

It was Cayce who brought castor oil packs to fame in the 20th century. The oil has a long and varied history of use as a healing agent in folk medicine around the world. According to a research report in a recent issue of the *Journal of Naturopathic Medicine*, castor bean seeds, believed to be 4,000 years old, have been found in Egyptian tombs, and historical records reveal the medicinal use of castor oil in Egypt (for eye irritations), India, China (for induction of childbirth and expulsion of the placenta), Persia (for epilepsy), Africa, Greece, Rome, Southern Europe, and the Americas. In ancient Rome, the castor oil plant was known as Palma Christi, which translates into *hand of Christ*. This name is still sometimes used today.

Edgar Cayce and the Palma Christi is a study of the use of castor oil and castor oil packs as suggested through Edgar Cayce and in observed cases followed in the practice of general medicine. In his book, over 50 different conditions of illness of the body are discussed where castor oil packs were the major or minor part of the therapy suggested.

A great many uses for castor oil are because of its lymph stimulating properties. Castor oil is recommended for use in massage, including applications for callouses, bunions, moles, warts, tumors, cancer (skin and breast), cysts, and many other conditions.

The use of castor oil in the form of packs was indicated for a number of conditions such as constipation, various liver conditions, appendicitis, arthritis, colitis, intestinal disorders and toxemia. Records of these uses of castor oil are located at the Edgar Cayce Foundation. The foundation is chartered in the state of Virginia and its primary function is the care of the 14,238 readings given by Edgar Cayce over a period of 40 years.

Dr. McGarey Interview:

The A.R.E. Clinic (Association for Research and Enlightenment), is a non-profit medical complex which offers a new holistic approach to health and healing. The philosophy of the clinic contains the concepts as found in the Edgar Cayce readings plus conventional medicine to provide a new approach to health.

The A.R.E. Clinic was founded by William McGarey, M.D., and his wife, Gladys McGarey, M.D. William McGarey graduated from the University of Cincinnati College of Medicine and Gladys McGarey from the Women's Medical College in Philadelphia.

Q. How did you get involved with Edgar Cayce's work?

McGarey: We began investigating his work and these healing phenomenon in 1955. In 1956, I personally met Hugh Lynn Cayce, Edgar Cayce's son, when he visited and spoke in Phoenix. From that point on my destiny was focused on Cayce's work linked with the A.R.E.

Q: What work do you include in the clinic that is associated with Edgar Cayce?

McGarey: Our whole attitude and philosophy behind the clinic is associated, but we also use some of the techniques he has mentioned. For example, we use castor oil remedies and castor oil packs a lot because of their affects on the body. In fact, not a day goes by that we do not recommend them. We have a research department where we are doing research in this particular area.

Q: Would you like to share some of the information you have found concerning castor oil?

McGarey: You can use castor oil packs on any part of the body and we do. These help the lymphatic system and the circulation of fluids through the system and body. It apparently has an anti.inflammatory affect. Cayce tells us when castor oil is applied, it acts to coordinate the activity between the functioning of organs in the body.

Castor Oil: What Is It?

Castor oil is derived from the oil from the bean-like seeds of the castor oil plant. The castor oil plant is a tropical plant (native to Africa), whose botanical name is ricinus communis, also referred to as the *Palma Christi,* and is noted as an important medicinal plant. The plant grows to heights of 8 to 12 feet with green or reddish 7 pointed star-like leaves. The blooms are reddish in color and the plant can be quite striking.

Scientific Name: *Ricinus communis*
Family: *Euphorbiaceae*

The castor bean has been known to man for thousands of years. Ancient Indians called it eranda, a name that is still used in some parts of Asia. It is also known to be a highly valued plant because seeds have been found in 4,000 year-old Egyptian tombs (Kowalchick, Hylton).

The ancient Greeks were also familiar with the castor bean and knew how to extract the oil which was only to be used externally. In the middle ages, herbalists in Europe used the oil externally as a kind of liniment and lubricant. Not until the 18th Century did it gain its reputation as a laxative (Kowalchick, Hylton) when taken internally.

The oil from the seeds is a colorless or yellowish oil historically credited as a cathartic (internal cleanser) and lubricant. The healing properties of castor oil have been known for the past several thousand years and treatments are likely to be found in ancient legends, home remedies and folk medicine.

Although many doctors reject folk remedies and herbal treatment as unscientific, they should not forget that the "root" of modern medicine grew out of our ancestors' appreciation for the very real healing properties of plants. Many well.known drugs today were once derived from barks or leaves, including: Aspirin derived from willow bark, ephedrine derived from ephedra herb, digitalis derived from foxglove, quinidine derived from cinchona bark, theophylline derived from tea leaves, and codeine derived from poppy. Many other medications, and many still in use today, derive their active ingredients from plants.

Castor Oil: Its Toxicity

Castor oil is obtained from the seeds or beans of the plant, but the rest of the seed and plant is considered to be extremely poisonous. Because of this, the plant is not bothered by insects and pests much so the foliage usually looks very nice. The plant and beans are poisonous to humans, pets, and other animals. It is even used as a natural repellent to moles and other similar animals which can be very destructive.

Castor oil, although known as a cathartic, is not recommended for internal use, because the seeds contain ricin, which is poisonous. Other parts of the plant, when just touched, are irritating to sensitive individuals. Ricin is a toxic protein present in the seed which coagulates red blood corpuscles. Consumed, ricin causes vomiting, diarrhea, thirst, and blurred vision. It only takes one of these poisonous seeds to kill a child (Kowalchick, Hylton). But, when extracted commercially, the oil is safe because the poisons are not soluble in oil and remain in the pomace when the seed is pressed. For this reason, only manufactured products are recommended.

Cayce, in addition, rarely suggested using castor oil as a cathartic (taken internally as a cleanser), and in some cases strongly advised against it.

Clinical signs of toxicity: Oral irritation, burning of mouth and throat, increase in thirst, vomiting, diarrhea, kidney failure, convulsions.

Castor Oil: What It Can Do

The Healing Properties of Castor Oil

Over 50 different conditions are recorded as successfully treated therapeutically with castor oil. Successful applications of castor oil have been recorded in disturbances of the digestive system including stomach, intestinal or colon problems; kidney, liver or gall bladder problems; disturbances of the lymphatic system; urinary and excretory systems; circulatory system; and some aspects of the nervous system.

McGarey lists many topical uses for castor oil. Warts, "liver spots," papillomas, and pigmented moles can be cleared up, any type of body ulcer and slow-to-heal umbilicus of a newborn infant can be healed, and can also be used for eye irritations, for lack of proper growth of hair in little children, for eyelashes or eyebrows to stimulate growth. Many examples of healing in each of these situations are cited.

Aching feet can be relieved of soreness and be made to feel much better if twice a week, or more often, they are rubbed down at bedtime with castor oil. Also, castor oil can be used to soften corns and calluses and remove the soreness. Castor oil can be massaged into the corns on feet both night and morning.

Although all this may sound a little corny and even "too good to be true," as one old family doctor said in 1965, "Castor oil will leave the body in better condition. tion than it found it."

The information included in the remainder of this

chapter is taken from the book, *Edgar Cayce: Encyclopedia of Healing*, by Reba Ann Karp. Following are examples of Cayce's recommendations for treatment of specific conditions where castor oil is mentioned as a major or minor part of the suggestion. If you are interested in following any of these suggestions, please consult a medical doctor. The information presented here is not recommended as a guide for self-healing.

Appendicitis

Appendicitis, inflammation of the appendix is often experienced in its severe state with rapid onset of its symptoms. Symptoms include pain in the right lower abdomen, anorexia and nausea, low fever and elevated white blood cell count. Poor elimination leading to build-up of toxins was mentioned as a factor for a majority of the cases Cayce saw. Poor circulation and spinal subluxation were the two other most frequent causes mentioned, and in many cases, a combination of two or more of the factors was involved.

Treatments centered on methods of increasing elimination, correcting diet, applying packs and administering massage and spinal manipulation. Castor oil packs were the most frequently suggested type of pack. Their "purpose was to relieve congestion in the entire gastrointestinal area and act as a counter-irritant to the properties added internally."

Castor oil packs may be applied without a heating pad if heat is felt to be undesirable.

Arthritis

Arthritis usually refers to-painful inflammation of the joints, but can involve any one of a hundred different types of arthritic diseases that affect the entire musculo-skeletal system. Cayce's many treatments involved the seven categories of causes of arthriti;, poor assimilation and elimination, impaired circulation, glandular malfunction, karmic and psychological causes, previous treatment, spinal subluxation and injuries.

Cayce's treatments were designed to restore the body's ability to function normally and stimulate the natural efforts of the system to heal itself. To relieve pain in afflicted joints for rheumatoid arthritis, local application of castor oil packs was one recommendation.

Cirrhosis of Liver

Cirrhosis is a chronic, progressive disease of the liver. Cayce cited poor elimination as a common cause of cirrhosis, toxemia and anemia were also mentioned as causes. Cayce's recommendations were based on improving eliminations and stimulating the liver, centering on the use of castor oil packs, doses of olive oil colonics or enemas, and proper diet. For example, in one case, castor oil packs were recommended two evenings a week for at least an hour or two, upon retiring, for about two weeks. After the third pack was applied, a dosage of olive oil was to be taken.

Colds

Treatments usually varied according to the cause. Changes in diet was often the first suggestion accompanied by drinking large amounts of pure water to flush the toxins out of the system. Massages, rubs, aids to elimination, spinal manipulation were also common recommendations. Application of various packs, such as castor oil, to the abdominal region, followed by a baking-soda water rinse, were also recommended.

Colic

Colic, is abdominal pain, usually found in infants under three months old. It is often accompanied by abdominal distention and large amounts of gas. Discomfort usually causes extended periods of crying. The most frequent cause of colic is poor assimilation due to improper diet. In the case of breast-fed babies, the mother's diet was at fault.

Successful use of abdominal castor oil packs for infants suffering with colic has been recognized for generations.

Colitis

A cold is an upper respiratory viral infection. Primary causes were infections, congestion, and bronchial inflammation, poor elimination leading to build-up of toxins in the system, dietary imbalances and over taxation of the nervous system and entire body.

Colitis is an inflammation of the colon causing the body to lose its ability to assimilate its foods properly.

Castor oil packs were recommended as well as dietary restrictions, spinal manipulations, and herbal tonics to normalize the intestinal tract.

Constipation

Constipation is often caused by blockage in the intestines, poor circulation in the intestines, spinal subluxation or poor diet and can lead to much more serious problems like toxicity, diverticulosis or tumors.

Cayce recommended gentle stimulation, spinal manipulation, and also frequently placing castor oil packs over the right side and also over the abdominal area to relieve lower bowel tension followed by a small dose of olive oil. Drinking large amounts of water also was suggested.

Disorders of the gallbladder, including gallstones, are the result of cholesterol formation or cholesteric crystals and other elements found in bile. These disorders are usually caused by poor digestive processes and poor elimination.

Recommended by Cayce to stimulate eliminations were small doses of olive oil taken at frequent intervals throughout the day coordinated with the application of castor oil packs several times a week.

Foot Problems

CALLUSES: A callus is a thickened, hardened area of the skin, usually on the bottoms of the feet or the palms of hands. Cause was attributed to spinal subluxation or impingement, poor circulation of lower limbs, or tight-fitting shoes.

Treatment to soften the callus usually included a topical solution such as massaging castor oil into the area.

CORNS: A corn is a cone-shaped horny mass of thickened skin on the toes which presses down on the skin beneath, making it thin and tender. Corns are usually the result of long-term friction and pressure, for example, from improper fitting shoes. Castor oil can be used to soften corns and remove the soreness. Castor oil can be massaged into the corns on feet both night and morning.

Gastritis and Indigestion

Indigestion, failure to properly digest food, includes symptoms of nausea, heartburn, flatulence, cramps, vomiting, and diarrhea. Gastritis is an inflammation of the stomach lining and can include any number of symptoms. Improper elimination, excess toxins, poor assimilation, spinal subluxation, and glandular imbalances were cited as the major causes of these digestive disorders.

Dietary advice was the major treatment suggested

in all cases. Abdominal packs were suggested in a number of cases, with castor oil packs the most frequently recommended.

Hepatitis

Hepatitis, inflammation of the liver caused by an infection, can be transmitted though various ways such as blood transfusions or contaminated food. Other causes, such as poor eliminations, resulting in torpid liver, were also listed by Cayce.

The most common treatments involved stimulants for eliminations and various packs intending to cleanse the liver of excess toxins. Castor oil packs to be placed over the abdomen and kidneys were recommended the most frequently.

Joint Stiffness

Joint stiffness is often caused by stagnation of fluids in the joint, sedimentation of minerals and irritation caused by uric acid. Uric acid is also associated with pain ful gout symptoms.

Joint stiffness can be relieved by massaging castor oil into the area. Relief occurs as circulation improves and the joint area is cleansed and soothed. Also, because castor oil helps to remove metabolic waste from the lymph, the blood is cleaner and uric acid and other materials will not build up in these areas.

Kidney Stones

Kidney stones are accumulations in the kidney resulting from accumulated deposits. When the stones become too large to pass through and out, they cause extreme pain and irritation.

Cayce believed kidney stones were caused by incoordination of circulation between the kidneys, liver and other related organs. Spinal subluxation was also attributed to contributing to the formation of kidney stones.

The treatments recommended by Cayce centered on decreasing inflammation and disintegrating the stone sufficiently enough for it to pass. Dietary changes, spinal manipulations, bedtime massages and also administration of castor oil packs are recommended.

Lesions

Lesion refers to any visible change in the tissue of the skin, such as a wound or injury, sore, rash, or boil.

Apply castor oil packs over the area until the lesions are broken up. The packs should be taken by periods, three days at a time, an hour each day. Follow this with two teaspoons of olive oil. After the castor oil packs have been applied for three days, skip a week and apply again.

Migraine Headaches

Migraine headaches are usually associated with severe tensions, creating exquisite pain and a number of other disturbances. Cayce regarded migraines as only a symptom of other internal imbalances. Causes listed were spinal misalignments causing nervous imbalances, poor elimination, and poor circulation causing build-up of toxins.

Cayce stressed the recommended treatments be applied in a consistent and persistent manner in order to restore a balance in the chemical reactions of the system. Dietary improvements to build nerves and blood, spinal manipulation, electrotherapy, and gentle laxatives were advised in many cases. Heated castor oil packs placed over the abdomen was an additional suggestion.

Moles and Warts

Moles and warts are skin elevations which appear as dark spots or skin colored spots. Warts are caused by viruses while moles have other causes, such as buildup of toxins.

To remove these skin elevations, Cayce frequently suggested massaging castor oil into the affected area. Castor oil was either to be used alone or mixed with baking soda.

Liver spots and papillomas can also be effectively treated by massaging castor oil into affected areas.

Neuritis

Neuritis is a painful inflammation of the nerves which can progress to severe pain and muscular weakness. Cayce listed poor eliminations and build-up of toxins as the cause for the majority of cases.

Treatment centered on elimination of accumulated toxins and dietary advice, but castor oil packs were also advised in some cases.

Psoriasis

Psoriasis is a skin condition where bright red lesions are covered by dry scales. The primary cause in each case was improper eliminations and build up of toxins, and therefore, treatment methods were focused on removing them and restoring the intestinal walls to normal. Dietary recommendations were focused on eating primarily alkaline foods (almonds, vegetables and fruit). Topical application of castor oil was mentioned, as well as dosages of olive oil.

Sluggish Liver

The liver has over 500 functions and the importance of maintaining optimal activity is crucial.

To stimulate liver activity, Cayce recommends application of castor oil over the liver area about one hour for two days and then taking two teaspoons of olive oil internally after the second day. Apply the castor oil pack with at least three thicknesses of flannel.

Toxemia

Toxemia occurs in the body when its systems can no longer efficiently remove the amount of toxins and waste produced by all of the body's metabolic processes. These toxins accumulate in the body from the estimated 13 to 17 pounds of dyes, preservatives, and pesticides an average person consumes in a year. Chemicals are also present in the water we drink and air we breath.

The liver is the major organ in the body responsible for detoxifying and eliminating substances no longer of use to the body. The following are Cayce's recommendations for a sluggish liver.

- Eat only small portions of meat and preferably fowl, fish or lamb.
- Eat a diet high in roughage, and high complex carbohydrates.
- Avoid sugars, sweets, and refined carbohydrates.
- Avoid fried and fatty foods.

Ulcers

Duodenal (peptic) ulcers are superficial sores on the mucous membranes which are usually on the first part (duodenum) of the small intestine leaving the stomach.

Poor circulation, emotional stress, poor elimination, and colds were cited as the main causes. Treatment centered on easing stress on the stomach (alkalizing), and stimulating eliminations. Suggested was drinking large amounts of water and use of castor oil packs.

Castor Oil and Olive Oil

In addition to recommending castor oil packs, general recommendations applicable to many, many conditions, but also for maintaining good health include:

- Maintain a proper alkaline/acid balance.
- Eat mainly fruits, vegetables, whole grains and only small portions of meat.
- Avoid red meat, especially pork and beef.

Cayce often recommends olive oil taken internally, in conjunction with the use of castor oil packs. Olive oil, a light oil pressed from ripe olives, has been medically recognized as a treatment for gallstones and also intestinal worms. Historically, olive oil was also known for a variety of uses.

Internally, olive oil is recommended as a laxative because it was considered a valuable food for the intestinal system, for inflammation of the gallbladder, toxemia, intestinal and liver disorders, and acidity.

Externally, according to Cayce, olive oil not only softens the skin, but is one of the most effective agents for stimulating muscular activity, or mucous membrane activity that can be applied to the body.

For these conditions, especially constipation, dosages of olive oil were to accompany use of castor oil packs.

Castor Oil: How To Apply

Castor oil has numerous external applications, which can be applied in two ways: By massaging the oil into the skin, or by making and applying castor oil packs.

Generally, castor oil packs were to be applied over the right abdomen according to specific cycles, such as three days of use alternated by three days of rest.

Instructions For Use of Castor Oil Packs

- Prepare a soft flannel cloth (wool is acceptable) folded two to three times thick. When folded it should be large enough to cover the entire area it is to be applied. For example, 8 x 12 inches for abdominal applications.
- Pour the castor oil in a pan and soak the cloth in the oil.
- Heat the oil-soaked cloth in the pan to as warm as the body can tolerate.
- Wring out the excess oil so it is not dripping wet and apply the cloth to the area on the body requiring treatment. (Bedding should be protected in some way to prevent soiling.)
- Cover the oil-soaked cloth with a plastic covering and then place a heating pad on the top. Begin the heat setting at medium and increase it to high if can be tolerated to keep the pack warm. It may be helpful to

wrap a towel around the entire area.

- Lukewarm soda water can be used to cleanse the area afterwards if desired.

- Store the flannel pack in the pan for future use.

- In addition, Cayce often recommended a small oral dose of olive oil on the third day of each cycle.

Topical Uses

Topical uses require the castor oil to be massaged into the skin. If the area cannot be wrapped or covered, excess oil can be wiped off with a tissue. Cover feet with cotton or wool socks, hands with cotton gloves, etc.

- The pack should remain in place between one an one-and-one-half hours.

- Frequency of use depends on the individual situation.

One recommendation is to apply the castor oil packs (before retiring) by periods of three consecutive days with a couple of days in between. Continue this pattern until results are seen. Another recommendation was to apply the pack every other evening at bedtime for 10 or more applications.

Castor Oil: How It Works

The body responds to castor oil through several means, with one of the primary impacts involving the lymphatic system. Parts of the nervous system, the liver, and entire digestive system are also affected by castor oil. Any activity of the body which is associated with the lymphatic system can benefit in some way.

Lymphatic System

Castor oil increases the activity and movement of lymph fluid through the vessels. Lymph vessels are located throughout the body and actually more numerous than blood vessels. Lymph is the inner excretory mechanism of the body. A vital function of the lymph system is cleansing metabolic waste and detoxifying the body by moving these through the system and out of the body. The lymphatic system provides the means through which each individual cell in the body can get rid of waste. Substances which are the result of cellular metabolism and which are extruded in the cell must be removed through the lymph. Lymph and the lymphatic vessels comes into a much more intimate relationship with the metabolic tissues than the blood does (McGarey).

Increasing the activity of the lymph through the body is of great importance. By increasing the metabolism of the system, fatty and other waste deposits are eliminated more quickly. Circulation of lymph is encouraged by physical activity, and also other things such as application of castor oil.

According to James A. Rick, wellness counselor and postural therapist, breathing and deep breathing are also very important in the circulation of lymph. Adequate circulation is a common problem for many people. Stagnant pools of lymph fluid are often located in the abdominal region partly because of the large amount of lymphatic vessels located there. Often this can occur because of poor posture and ineffective breathing patterns.

Deep abdominal breathing is rare in Western culture. Such breathing acts as a pump for the diaphragm causing contraction and expansion. This not only stimulates movement of abdominal lymph but also massages the organs beneath this area.

Nervous System

The functioning of the lymphatic system is closely allied with the normal functioning of the autonomic nervous system (involuntary reactions such as digestive and activity of the heart). Part of the autonomic nervous system, the parasympathetic, is said to be is responsible for the rebuilding and healing of the body.

According to McGarey, castor oil, absorbed in the tissues, may, in its vibratory activity act to stimulate the parasympathetic nerve supply which is located in the area being treated. This, in turn, would stimulate the lymphatics to drain more adequately the tissues under stress. This is beneficial to any organ or portion of the body which is clogged with waste products.

Another function of the castor oil pack is enhancement of the liver function, "not only in its ability to be the great detoxifying organ of the body, but also in its beneficial effect to all the surrounding organs rather than being as a dross and a distress to them. In some way, Cayce implies, when the liver is not functioning normally, it can and often does act as an irritant to some or all of the organs which surround it in the abdominal and the chest cavity."

The liver produces one-third to one-half of all the lymph produced in the human body under resting conditions. This lymph, along with the lymph from the intestines, constitutes half of all the lymph produced in the body.

Digestive System

The digestive system is responsible for ingestion, digestion, assimilation, and elimination. Improper elimination and poor dietary habits are the major contributors to most digestive disorders.

Castor oil acts as a stimulant for lymph and blood, improving all digestive mechanisms. The lymphatic system is one of the major channels of absorption from the gastrointestinal tract, being responsible for the absorption of fats. In addition to the responsibility of the lymph system to cleanse cells, the lacteals are associated with the absorption and assimilation of foods.

Bibliography

Karp, Reba Ann, "Edgar Cayce, Encyclopedia of Healing" (1986) Warner Books, New York, N.Y

Kowalchick, Claire, and William Hylton, "Rodale's Illustrated Encyclopedia of Herbs" (1987) Rodale Press, Emmaus, Pennsylvania.

McGarey, William, M.D., "Edgar Cayce and the Palma Christi," (1970) Edgar Cayce Foundation, Virginia Beach, Virginia.

ABOUT THE AUTHOR

Beth M. Ley, Ph.D., has been a science writer specializing in health and nutrition since 1988 and has written many health-related books, including the best sellers, ***DHEA: Unlocking the Secrets to the Fountain of Youth*** and ***MSM: On Our Way Back to Health With Sulfur***. She wrote her own undergraduate degree program and graduated in Scientific and Technical Writing from North Dakota State University in 1987 (combination of Zoology and Journalism). Beth has her masters (1998) and doctoral degrees (1999) in Nutrition.

Beth does nutrition and wellness counseling at The Wellness Center in Detroit Lakes, MN, and speaks on nutrition, health and divine healing locally and nationwide.

Beth lives in the Minnesota lakes country. She is dedicated to God and to spreading the health message. She enjoys nature and spending time with her dalmatian, KC.

Memberships: American Academy of Anti-aging, New York Academy of Sciences, Oxygen Society and Resurrection Apostolic International Network (RAIN), Gospel Crusade.

Other Recommended Titles by Beth Ley, Ph.D.

ISBN: 1-890766-19-4
240 pages, $14.95

ISBN: 0-9642703-0-7
110 pages, $8.95

Credit card orders call toll free:
1-877-BOOKS11

Also visit:
www.blpublications.com

Books from BL Publications

Calcium: The Facts
Beth M. Ley, Ph.D., 56 pgs, $4.95

Calcium is important for health of the heart, bones and teeth, blood pressure, and needs to be in balance with other minerals in the body. Find out the best sources - such as coral calcium - and not so good sources - daily products! and much more! Minerals help keep the body slightly alkaline... an acidic system from refined foods opens the door for sickness and disease!

Flax! Fabulous Flax!
Beth M. Ley, Ph.D. 40 pgs. $4.95

Flax is the best source of essential Omega-3 fatty acids and lignin fiber. These help inflammatory disorders, help regulate cholesterol levels, glucose levels, blood pressure, acne, psoriasis & other skin problems, protect against cancer, heart disease, diabetes & most other degenerative conditions!

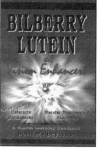

Bilberry & Lutein: The Vision Enhancers
Beth M. Ley, Ph.D. 40 pgs. $4.95

Natural protection against free radical attack which disturbs vision health resulting in cataracts, macular degeneration glaucoma, retinal diseases and others!

Discover the Beta Glucan Secret!
Beth M. Ley, Ph.D., 2001, 40 pages, $3.95

Beta Glucan, a natural component of yeast, triggers an immune response in the body creating a system of defense against viral, bacterial, fungal, parasitic or potentially cancerous invaders. Research shows its excellent benefits for immune enhancement, cancer prevention & treatment, cholesterol reduction, glucose regulation and much more!

Aspirin Alternatives:
The Top Natural Pain-Relieving Analgesics
Raymond Lombardi, D.C., N.D., C.C.N., 1999, 160 pages, $8.95

This book discusses analgesics and natural approaches to pain. Ibuprofen and acetaminophen are used for pain-relief, but like all drugs, there is a risk of side effects and interactions, Natural alternatives are equally effective and in many cases preferable because they may help treat the underlining problem rather than simply treating a symptom.

Diabetes to Wholeness!
Beth M. Ley, 120 pgs. $9.95

A natural and spiritual approach to disease prevention and control. Covers TypeI and Type II. Learn about significance of whole foods, protective supplements and spiritual roots.

MSM: On Our Way Back To Health With Sulfur
Beth M. Ley, 40 pages, $3.95

MSM (methyl sulfonyl methane), is a rich source of organic sulfur, important for connective tissue regeneration. Beneficial for arthritis and other joint problems, allergies, asthma, skin problems, TMJ, periodontal conditions, pain relief, and much more! Includes important "How to use" directions.

How to Fight Osteoporosis & Win: The Miracle of MCHC
Beth M. Ley, 80 pgs. $6.95

Find out if you are at risk for osteoporosis and what to do to prevent and reverse it. Get the truth about bone loss, calcium, supplements, foods, MCHC & much more!

Coenzyme Q10: All Around Nutrient for All-Around Health!
Beth M. Ley-Jacobs, 1999, 60 pages, $4.95

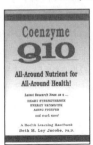

CoQ10 is found in every living cell. With age, insufficient levels become more common, putting us at serious risk of illness and disease. Protect and strengthen the cardiovascular system; benefit blood pressure, immunity, fatigue, weight problems, Alzheimer's, Parkinson's, Huntington's, gum-disease and slow aging.

DHA: The Magnificent Marine Oil
Beth M. Ley-Jacobs, 1999, 120 pages, $6.95

Individuals commonly lack this essential Omega-3 fatty acid so important to the brain, vision, and immune system and much more. Memory, depression, ADD, addiction disorders, inflammatory disorders, skin problems, schizophrenia, elevated blood lipids, etc., benefit from DHA.

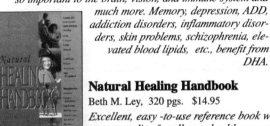

Natural Healing Handbook
Beth M. Ley, 320 pgs. $14.95

Excellent, easy-to-use reference book with natural health care remedies for all your healthcare concerns. A book you will use over & over again!

# of copies	

TO PLACE AN ORDER:

- ___ *Aspirin Alternatives: The Top Natural Pain-Relieving Analgesics* (Lombardi) $8.95
- ___ *Bilberry & Lutein: The Vision Enhancers!* (Ley) $4.95
- ___ *Calcium: The Facts, Fossilized Coral* (Ley) $4.95
- ___ *Castor Oil: Its Healing Properties* (Ley) $3.95
- ___ *Dr. John Willard on Catalyst Altered Water* (Ley) $3.95
- ___ *Chlorella: Ultimate Green Food (Ley)* $4.95
- ___ *CoQ10: All-Around Nutrient for All-Around Health* (Ley) $4.95
- ___ *Colostrum: Nature's Gift to the Immune System* (Ley) $5.95
- ___ *DHA: The Magnificent Marine Oil* (Ley) $6.95
- ___ *DHEA: Unlocking the Secrets/Fountain of Youth-2nd ed.* (Ash & Ley) ...$14.95
- ___ *Diabetes to Wholeness* (Ley) $9.95
- ___ *Discover the Beta Glucan Secret* (Ley) $3.95
- ___ *Fading: One family's journey ... Alzheimer's* (Kraft) $12.95
- ___ *Flax! Fabulous Flax!* (Ley) $4.95
- ___ *Flax Lignans: Fifty Years to Harvest* (Sönju & Ley) $4.95
- ___ *God Wants You Well* (Ley) $14.95
- ___ *Health Benefits of Probiotics* (Dash) $4.95
- ___ *How Did We Get So Fat? 2nd Edition* (Susser & Ley) $8.95
- ___ *How to Fight Osteoporosis and Win!* (Ley) $6.95
- ___ *Maca: Adaptogen and Hormone Balancer (Ley)* $4.95
- ___ *Marvelous Memory Boosters* (Ley) $3.95
- ___ *Medicinal Mushrooms: Agaricus Blazei Murill (Ley)* $4.95
- ___ *MSM: On Our Way Back to Health W/ Sulfur* (Ley) SPANISH $3.95
- ___ *MSM: On Our Way Back to Health W/ Sulfur* (Ley) $3.95
- ___ *Natural Healing Handbook* (Ley) $14.95
- ___ *Nature's Road to Recovery: Nutritional Supplements for the Alcoholic & Chemical Dependent* (Ley) $5.95
- ___ *PhytoNutrients: Medicinal Nutrients in Foods, Revised /Updated* (Ley) $5.95
- ___ *Recipes For Life! (Spiral Bound Cookbook)* (Ley) $19.95
- ___ *Secrets the Oil Companies Don't Want You to Know* (LaPointe) ... $10.00
- ___ *Spewed! How to Cast Out Lukewarm Christianity through Fasting and a Fasted Lifestyle -* $15.95
- ___ *The Potato Antioxidant: Alpha Lipoic Acid* (Ley) $6.95
- ___ <u>*Vinpocetine: Revitalize Your Brain w/ Periwinkle Extract!* (Ley)</u> $4.95

Subtotal $ _____ Please add $5.00 for shipping **TOTAL $**_____

Send check or money order to:
BL Publications 649 Kayla Lane, Hanover, MN 55341
Credit card orders please call toll free: 1-877-BOOKS11
For more info visit: www.blpublications.com